MW00745106

THE LIBRARY OF THE PLANETS™

PLUTO

Luke Thompson

The Rosen Publishing Group's
PowerKids Press™
New York

Published in 2001 by The Rosen Publishing Group, Inc.
29 East 21st Street, New York, NY 10010

First Edition

Book Design: Michael Caroleo and Michael de Guzman

Photo Credits: pp. 1, 4, 20, 22 PhotoDisc; p. 7 (Roman god Pluto) Michael R. Whalen/NGS Image Collection, p. 7 (Pluto) PhotoDisc (digital illustration by Michael de Guzman); p. 8 (Sun) PhotoDisc, p. 8 (Pluto and Charon) Davis Meltzer/NGS Image Collection (digital illustration by Michael de Guzman); p. 11 © CORBIS; p. 12 Davis Meltzer/NGS Image Collection; p. 15 (Pluto) Ludek Pesek/NGS Image Collection, p. 15 (surface of Pluto) Courtesy of NASA/JPL/California Institute of Technology (digital illustration by Michael de Guzman); pp. 16, 19 Courtesy of NASA/JPL/California Institute of Technology.

Thompson, Luke.
Pluto/ by Luke Thompson.
p. cm.— (The library of the planets)
Includes index.
Summary: Examines the history, unique features, and exploration of the ninth planet from the Sun.
ISBN 0-8239-5650-4
1. Pluto (Planet)—Juvenile literature. [1. Pluto (Planet)] I. Title. II. Series.

QB701 .T46 2000
523.48'2–dc21 00-024766

Manufactured in the United States of America

Contents

1	The Ninth Planet	5
2	Discovering Pluto	6
3	Pluto's Orbit	9
4	A Double Planet?	10
5	Tilting Pluto	13
6	The Surface of Pluto	14
7	Changing Seasons	17
8	Pluto and Triton	18
9	Is Pluto a Planet?	21
10	A Mission to Pluto	22
	Glossary	23
	Index	24
	Web Sites	24

Charon

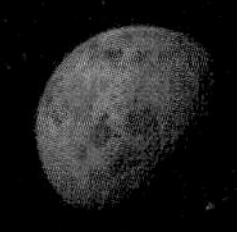

Pluto

The Ninth Planet

Of the nine planets in our **solar system**, Pluto is the farthest from the Sun. This is why it is called the ninth planet. Pluto is also the smallest planet in the solar system. It is only 1,440 miles (2,317 km) across. That means that it is about two-thirds the size of Earth's Moon.

Pluto's average distance from the Sun is 3.7 billion miles (5.9 billion km). Even when Pluto is close to the Sun, it is still almost 30 times farther away from the Sun than Earth is. In the year 2113, Pluto will be as far away from the Sun as it ever has been. At that point it will be almost 50 times farther from the Sun than Earth is.

This is an illustration of Pluto and its moon, Charon. The illustration was made by a computer that put together many images taken from a satellite, a human-made spacecraft that orbits planets.

Discovering Pluto

Pluto was discovered by an **astronomer** named Clyde Tombaugh in 1930. Scientists were looking for a planet that they thought was changing the **orbits** of two other planets, Neptune and Uranus. An orbit is the path a planet follows around the Sun. The scientists named the planet they were looking for Planet X. They had a clear idea of where they thought Planet X would be. It turned out that Planet X was too small to affect Neptune and Uranus. It was also too far away from them. Still, Tombaugh found this planet very close to where the scientists thought it would be. Planet X was named Pluto after the Greek god of the underworld. All the planets except Earth are named after gods in Greek or Roman mythology. The **ancient** Greeks believed that after you died, you went to the underworld. Some people think that Pluto got its name because it is a faraway, cold planet.

Pluto is named after the Greek god of the underworld. The "PL" of Pluto is the ninth planet's symbol.

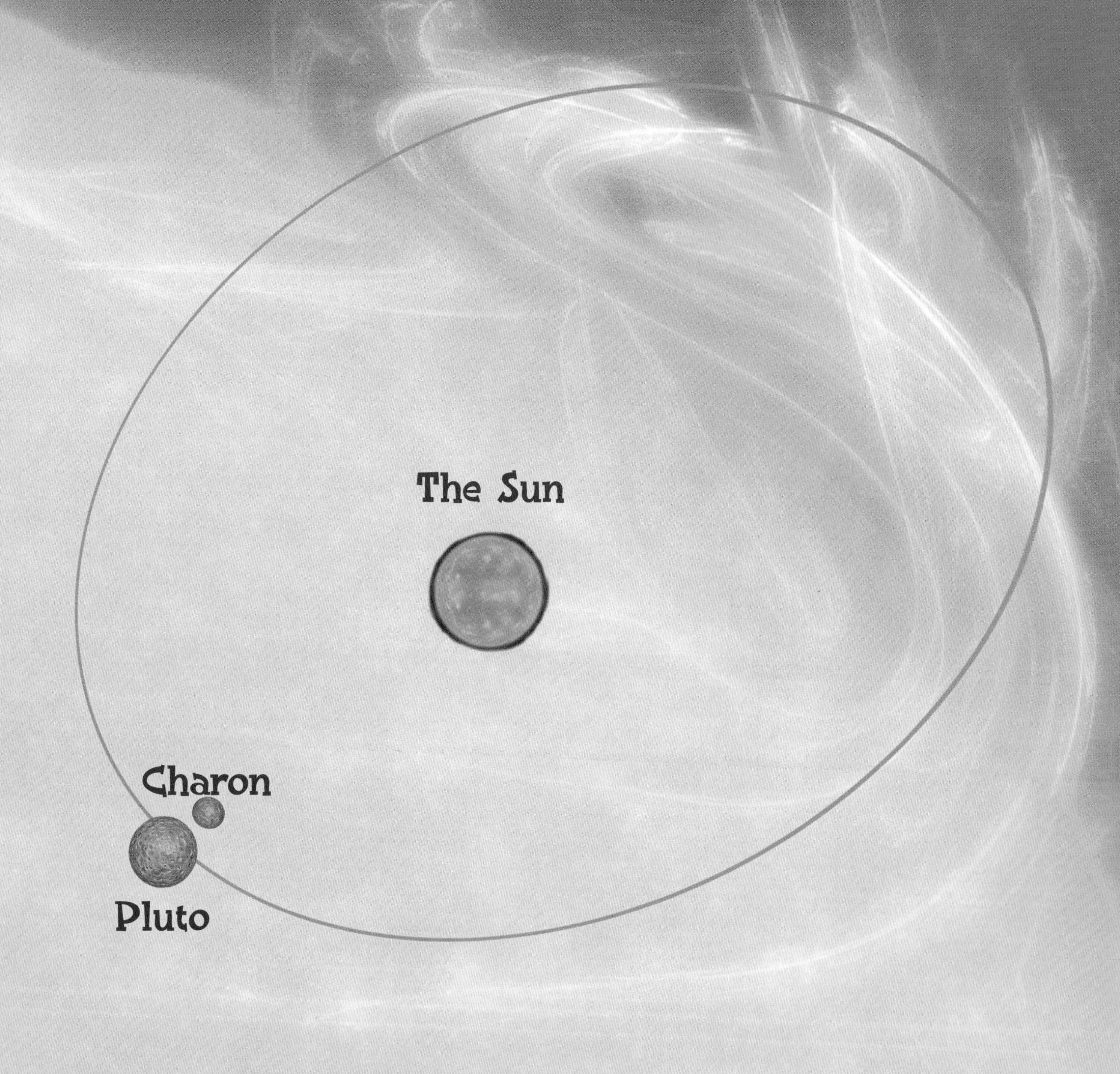
The Sun
Charon
Pluto

Pluto's Orbit

Most planets orbit the Sun in paths that look like stretched-out circles. Pluto's orbit is different. It is narrower and more stretched out. Pluto's strange orbit means that it is not always the same distance from the Sun. Pluto's orbit lasts 248 Earth years. For 20 years of its 248-year orbit, Pluto is closer to the Sun than Neptune is. During this time it is the eighth planet from the Sun. This last happened between 1979 and 1999. On February 11, 1999, Pluto crossed Neptune's orbit and became the ninth planet once again.

Planet	Orbit Time Around the Sun
Mercury	88 Earth days
Venus	225 Earth days
Earth	1 year (365 days)
Mars	1 year and 322 Earth days
Jupiter	12 Earth years
Saturn	29 Earth years
Uranus	84 Earth years
Neptune	165 Earth years
Pluto	249 Earth years

From 2226 until 2246, Pluto will be the eighth planet from the Sun once again.

A Double Planet?

So far, scientists have discovered only one moon for Pluto. That moon is called Charon. Charon is 790 miles (1,271 km) across, almost one-third the size of Pluto. No other planet has a moon that is so close to its own size. Charon is also very close to Pluto. It orbits Pluto at a distance of 12,000 miles (19,312 km). Our own Moon orbits Earth at a distance of 238,000 miles (383,024 km).

Some scientists think that Pluto and Charon are close in size because they are really a double planet. A double planet occurs when two bodies are close in size and orbit around each other. Charon does not just orbit Pluto. Pluto orbits Charon as well. This makes them different from other planets and their moons. If Pluto and Charon do make up a double planet, it is the only one we know of in the solar system.

In the ancient Greek religion, Charon was the boatman who took souls to the underworld.

CHARON

PLUTO

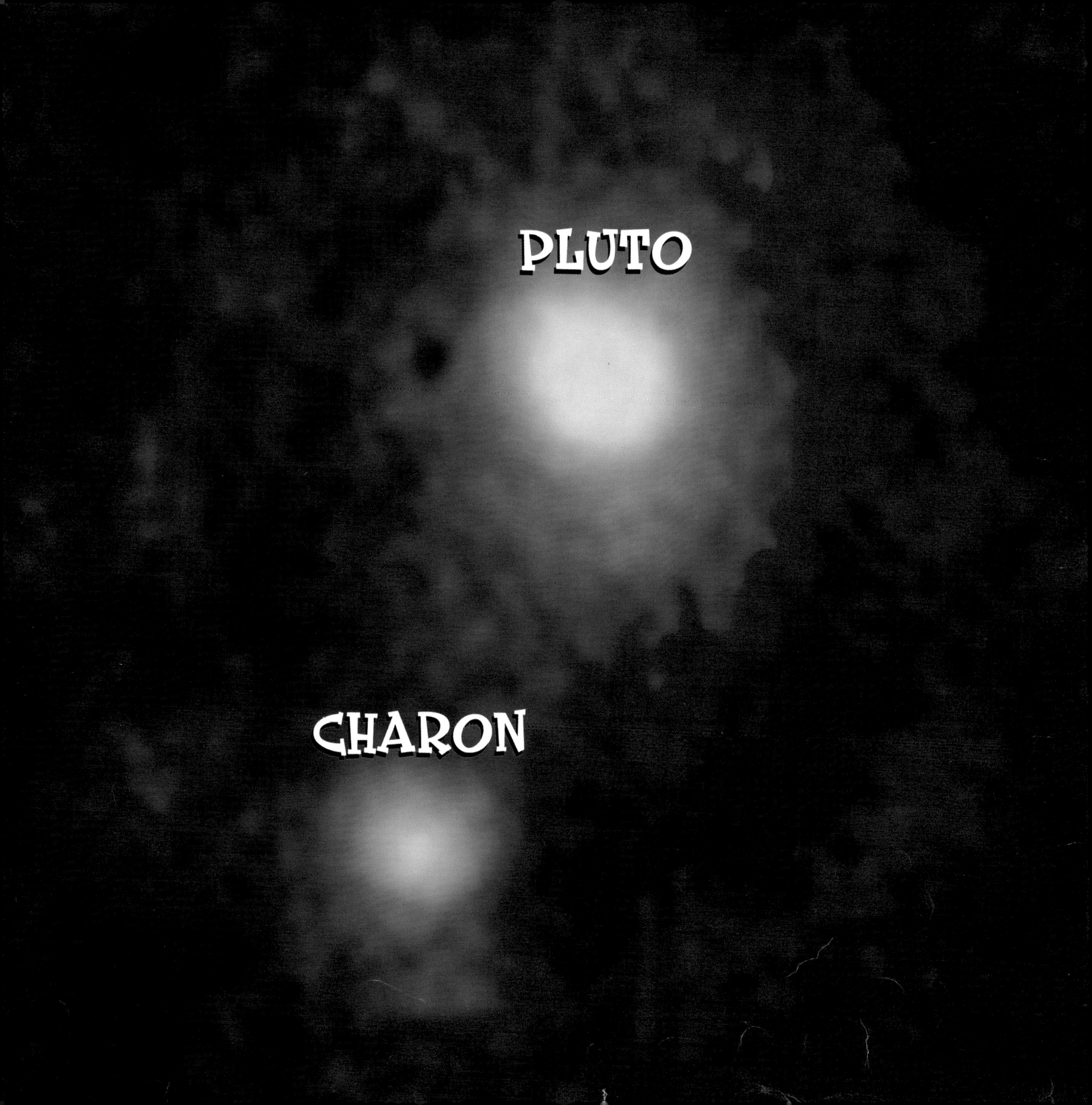

PLUTO

CHARON

Tilting Pluto

Like Uranus, Pluto is tilted on its side. Pluto completes half its orbit every 124 years. During this time, it looks like Charon is passing over and under Pluto. It looks like this for a period of about five years. The next time we will see Charon pass over and under Pluto will be in 2019. Scientists have learned a lot about Pluto during these five-year time periods. If Charon passes in front of Pluto and Pluto looks darker, scientists think that Charon is covering a bright spot on Pluto. If there is little change in Pluto's brightness, Charon is probably covering an area of Pluto that is darker.

This image of Pluto and Charon was taken by a camera on the Hubble Space Telescope. The Hubble Space Telescope is a powerful telescope that orbits Earth and sends back images of the solar system.

The Surface of Pluto

Pluto is a very cold planet. It has a surface temperature of about -450 degrees Fahrenheit (-268 degrees C). On Earth the lowest temperatures recorded are about -100 degrees Fahrenheit (-73.3 degrees C). Pluto is so cold that no form of life that we know of could survive on it. Scientists believe that Pluto is made up of 75 percent rock and 25 percent ice. Pluto's icy surface is mostly made up of the **gas nitrogen**. There are also small

This is a map of the surface of Pluto taken by the Hubble Space Telescope. Four different images were taken and then put together using a computer.

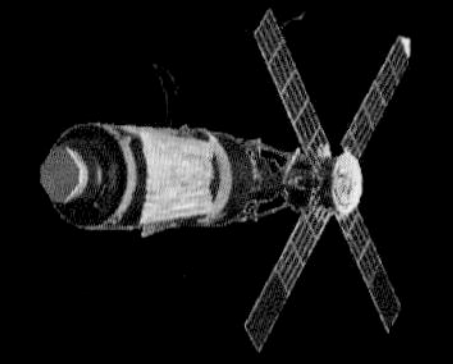

amounts of **methane** and **carbon monoxide**, two other gases. Pluto is so cold that these gases freeze at the planet's northern and southern **poles**. When Pluto moves closer to the Sun during its orbit, some of this ice melts and the gases are released into the **atmosphere**. The atmosphere is the layer of gases that surrounds a planet.

Changing Seasons

Even though Pluto is a small, cold planet, it does have changing seasons. These changes are based on Pluto's orbit. Sometimes Pluto is as close as 2.8 billion miles (4.5 billion km) from the Sun. Other times it is as far as 4.6 billion miles (7.4 billion km) from the Sun. As Pluto pulls away from the Sun, much of its atmosphere freezes into ice on the surface. When it gets warmer, the ice melts. A new layer of ice forms at the start of Pluto's 248-year orbit. The surface temperature on Pluto is now between -350 degrees Fahrenheit (-212 degrees C) and -380 degrees Fahrenheit (-229 degrees C). These differences in temperature set up differences in pressure that cause high winds. In its current orbit, Pluto passed closest to the Sun in 1989. The last time Pluto was this close to the Sun, George Washington was a little boy!

These images of Pluto were taken from the Hubble Space Telescope.

Pluto and Triton

Some scientists think that Pluto is more like Triton, Neptune's largest moon, than it is like any other planet. They think that Pluto and Triton might have been formed at the same time. This might explain why they are so similar. Triton is 1,700 miles (2,736 km) across, while Pluto is 1,400 miles (2,253 km) across. Triton and Pluto are both very cold. The temperature on Triton was once measured at -400 degrees Fahrenheit (-240 degrees C). Triton and Pluto are also both made of 75 percent rock and 25 percent ice. In 1977, The United States **launched** a **space probe** named *Voyager 2*. *Voyager 2* showed us that there were frozen lakes on Triton. There were also **geysers** on Triton that spit nitrogen gas five miles (8 km) into the sky. Pluto might have the same **geology** and weather as Triton. Until a space probe gets close to Pluto, though, we won't know for sure.

This picture of Triton's surface was taken by the Voyager 2 space probe.

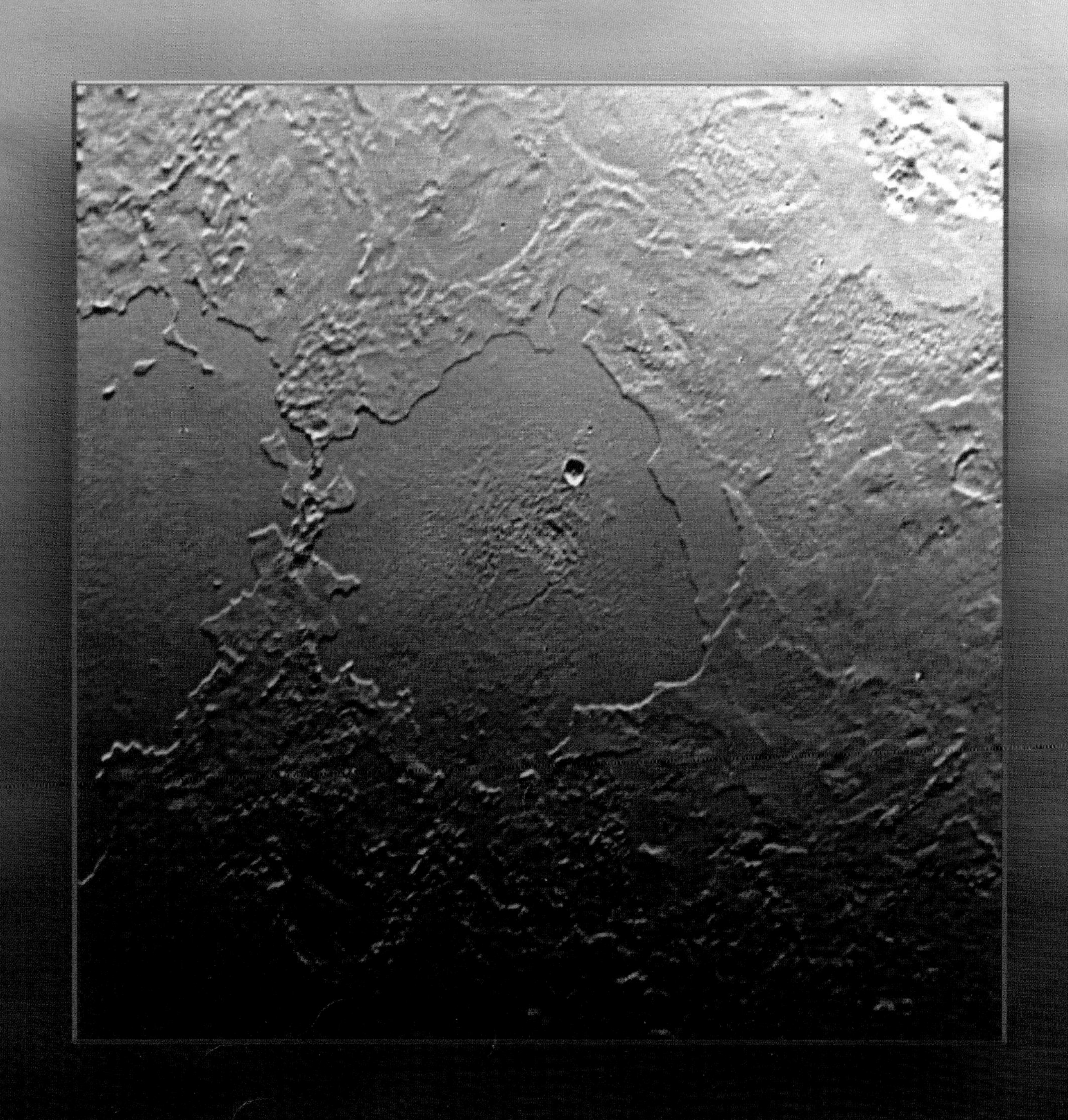

Is Pluto a Planet?

Some scientists think that Pluto is not really a planet at all. They say that it is too small and too different from the other eight planets. Scientists have also discovered thousands of large, icy bodies in orbits beyond Pluto. These icy bodies are in an area called the Kuiper Belt. They are so much like Pluto that they are called "Plutinos," which means "little Plutos." There are scientists who think that Pluto and its moon, Charon, were once two of these icy bodies. Maybe in years to come some of the icy bodies in the Kuiper Belt will be called planets. There is still much to discover about our solar system. One reason science is so exciting is that things are always changing!

Pluto and the other eight planets are part of our solar system, which is part of our galaxy, the Milky Way. Galaxies are large groups of stars and the planets that circle the stars.

A Mission to Pluto

Pluto is the only planet in our solar system that has not been seen by a space probe close up. A powerful **telescope** called the Hubble Space Telescope has sent back images from Pluto. Still, there is much to learn about Pluto and its moon, Charon. The United States is planning to launch a space probe called the *Pluto-Kuiper Express*. This space probe will reach Pluto and Charon in 2010. If that **mission** goes well, the space probe may go on to the Kuiper Belt. There it will study the large, icy bodies that have recently been discovered. As the farthest planet from the Sun, Pluto has been difficult to study. With the launch of the *Pluto-Kuiper Express*, we hope that will change. There are still so many interesting things to discover about this faraway, **fascinating** planet.

ancient (AYN-chent) Very old, from a long time ago.
astronomer (ah-STRAH-nuh-mer) A person who studies the sun, moon, planets, and stars.
atmosphere (AT-muh-sfeer) The layer of gases that surrounds an object in space. On Earth, this layer is air.
carbon minoxide (KAR-bin min-OK-syd) A colorless, odorless, very poisonous gas.
fascinating (FAS-in-ayt-ing) Very interesting.
gas (GAS) A substance that is not liquid or solid, has no shape or size of its own, and can expand without limit.
geology (jee-OL-oh-jee) The makeup of rocks and minerals on a certain surface.
geysers (GUY-serz) Springs that send up jets of hot water or gas.
launched (LAWNCHD) Pushed out or put into the air.
methane (MEH-thayn) A colorless, odorless gas.
mission (MISH-shun) A trip taken for a special purpose.
nitrogen (NY-troh-jen) A gas without taste, color, or odor that can be found in the air.
orbits (OR-bitz) The paths objects take around another object, such as the Sun.
poles (POHLS) The points that mark the ends of the straight line around which an object turns.
solar system (SOH-ler SIS-tem) A group of planets that circle a star. Our solar system has nine planets, which circle the Sun.
space probe (SPAYS PROHB) A spacecraft that travels in space and is steered by scientists on the ground.
telescope (TEL-uh-skohp) An instrument used to make distant objects appear closer and larger.

A
atmosphere, 15, 17

D
double planet, 10

G
gas, 14, 15, 18
geology, 18
geysers, 18

H
Hubble Space Telescope, 22

I
ice, 14, 15, 17, 18, 21, 22

K
Kuiper Belt, 21

M
moon, 5, 10, 21, 22

O
orbits, 6, 9, 10, 13, 15, 17, 21

P
Pluto-Kuiper Express, 22

R
rock, 14, 18

S
solar system, 5, 10, 21, 22
space probe, 18, 22

T
telescope, 22
temperature, 14, 17, 18
Tombaugh, Clyde, 6

V
Voyager 2, 18

To find out more about Pluto, check out these Web sites:
http://seds.lpl.arizona.edu/nineplanets/nineplanets/pluto.html
http://spaceplace.jpl.nasa.gov/site_index.htm